LI

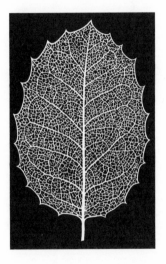

"Li reflect the order and pattern in Nature ... but it is not pattern thought of as something dead, like a mosaic: it is dynamic pattern as embodied in all things living, and in human relationships, and in the highest human values."

Joseph Needham, Science and Civilisation in China

Originally published in Wales by Wooden Books Ltd. in 2003;
first published in the United States of America in 2003 by
Walker Publishing Company, Inc.

Published simultaneously in Canada by Fitzhenry and Whiteside,
Markham, Ontario L3R 4T8

Printed on recycled paper

For information about permission to reproduce selections from this
book, write to Permissions, Walker & Company, 435 Hudson Street,
New York, New York 10014

Library of Congress Cataloging-in-Publication Data available upon request
ISBN 0-8027-1410-2

Visit Walker & Company's Web site at www.walkerbooks.com

Printed in the United States of America

2 4 6 8 10 9 7 5 3 1

LI

DYNAMIC FORM IN NATURE

written and illustrated by

David Wade

Walker & Company
New York

For Becky, Zeke, and Jesse

Any reader wanting to further investigate the mathematical aspects of dynamic forms in nature could do no better than to consult The Self-made Tapestry *by Phillip Ball, Oxford University Press, 1999. The image on the half title page shows a skeletonized holly leaf. The image on the title page is of paralleled geological folding. On this page we see the fir-tree domain pattern on the surface of a crystal (above), and the meandering jointing of the plates in a skull (below).*

CONTENTS

A positive Lichtenberg figure, obtained when a block of plastic is charged with static electricity.

INTRODUCTION

"There is a haunting if deceptive modernity in the notion, so often celebrated by baroque poets and thinkers, that arteries and the branches of trees, the dancing motion of the microcosm and the solemn measures of the spheres, the markings on the back of the tortoise and the veined patterns on rocks, are all ciphers."

George Steiner, Lifelines

The images presented in this book derive from a broad range of naturally occurring formations of a kind that have long been of interest and a source of inspiration to artists and designers but which have never been systematically investigated in the West and have not even acquired the dignity of a common term.

The expression *li*, which I have applied to these configurations, has been appropriated from the Chinese where it has been used for a very long time indeed. In common with many Chinese expressions, translations are various, but as a concept it falls between our notions of pattern and principle. It should become clear from even the most casual glance through the pages of this book that these formations have a certain universality, a quality that undoubtedly underlies their aesthetic appeal.

In fact *li* can be seen as a manifestation of the gestalt, the inherent pattern of things. In its earliest Chinese usage the term was applied to such phenomena as the markings in jade and the pattern of fibers in muscle, but it gradually acquired the more extended meaning of an innate principle.

What we are dealing with here then are graphic expressions of a great range of archetypal modes of action, the traces of which may be found throughout the natural world. They present, in a traditional Chinese view at least, an order that arises directly out of the nature of the universe. It is this somewhat Platonic notion of eternal and preexistent forms that thus accounts for the appearance of strikingly similar formations in widely different circumstances and in quite unrelated phenomena.

That an ancient Chinese term should be appropriate for this particular category of graphic imagery is itself part of that by now well-known convergence of older Eastern (and of mystical) concepts with more recent developments in our own Western, more analytical, traditions of thought.

Western science has always been interested in pattern, indeed pattern recognition can be seen as the very basis of science. But it is only in more recent times that the sort of quasi symmetrical forms that are presented here have been considered worthy of serious investigation. This has involved a great extension of symmetry concepts, and a moving away from rigid classicism. In fact physicists have long abandoned older notions of a static, material substantiality in favor of a view that regards substance as the product of energetic forces that are ceaselessly at work in the universe. In this, of course, the integrative quality of current scientific views brings them much more in line with older Eastern religious and philosophical concepts. There is an aspect of *li,* however, to which Western thought, at least on a general philosophical level, has not yet aspired; this concerns the notion of a phenomenal dualism, which is so much a part of Eastern (particularly Chinese) cosmological views.

Li are essentially dynamic formations, and as such can give the impression of a frozen moment, of a process caught at a particular instant of time, or, in a more abstract sense, of the principle of energy engaging with that of form. *Li* are appealing in a purely aesthetic sense because, although they tend to be relatively simple configurations, they have a high degree of content.

In this era of microprocessing we are all becoming familiar with the extent to which information of all kinds can be condensed and stored. Perhaps because of this the descriptive potentialities of the sorts of squiggles, blobs, and striations that are presented here are more easily appreciated than in even the more recent past. If this is indeed the case it would seem that the aesthetics of *li* are as characteristically modern as they are respectfully ancient.

According to the great Sung philosopher Chu Hsi (1130–1200 C.E.): "The term *tao* refers to the vast and great; the term *li* includes the innumerable veinlike patterns included in the Tao. . . . *Li* is like a piece of thread with its strands, or like this basket. One strip goes this way, and the other goes that way. It is also like the grain in bamboo. On the straight it is of one kind, and on the transverse it is of another kind. So also the mind possesses various principles."

Llanidloes, 2002.

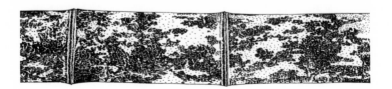

AGGREGATION
the collocation of elements

According to the early Greek philosopher Empedocles all the change and movement in the universe, including all the processes of creation and destruction, stem ultimately from two great principles of attraction and repulsion.

Clearly, form of any kind depends on its component parts holding together, just as its ultimate dissolution is a result of falling apart.

The *li* opposite, characteristic of the tenuous, temporary alliances seen on the surface of a liquid medium as particulate clustering (*opposite, top*) or suds (*opposite, bottom*), have only the minimal requirement for a recognizable structure. And yet form is apparent here, to an extent that is not discernible in, say, the swirls of an entirely liquid medium.

Structural formality becomes even more apparent when an aggregation forms around a nucleic center, particularly where, as in the examples of a soot cluster (*below, left*) and a bacterial growth pattern (*below, right*), an elementary branching formation appears.

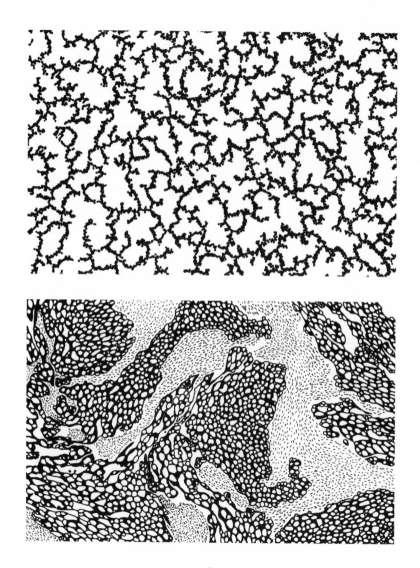

ANFRACTUOUS
winding and turning forms

The irregular streaming forms of the kind opposite derive from a number of different and quite unrelated phenomena.

The upper example is, in fact, an *equilibrium pattern* resulting from the Kerr magneto-optic effect (in a thin section of barium ferrite); that below is from a section of brain.

There are obvious points of similarity between these and the familiar whorls of fingerprints and the banding patterns in animal markings. In a similar way the suture lines of a fossilized ammonite shellfish (*below*) can also be envisaged as an aerial view of a meandering great river.

In common with many *li*, the processes by which these forms come into being can sometimes be traced back to structures associated with their underlying causation, and sometimes not; the reasons for this are still mysterious.

It is also the case that they may appear at every level of scale, from the micro- to the macroscopic, a fact that provides additional testimony to their essentially archetypal nature.

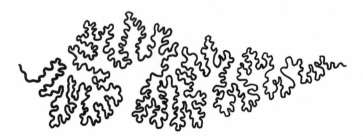

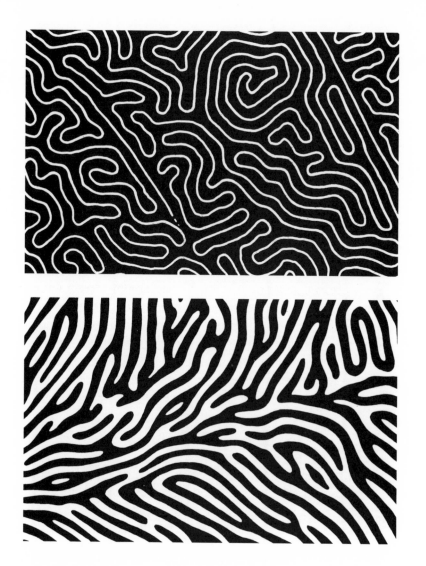

ANGULATED
formed with angles

The interlocking crystalline forms of river ice (*opposite, top*) pile onto each other to create the appearance of geometrical confusion. In fact there are vestiges of order here, but they are occluded.

The common crystalline habit of settled ice is expressed as hexagonal plates (a fact recorded by the Chinese in very early times). This regularity is obvious in snowflakes, but in freeze-thaw conditions a conglomeration of ice crystals of different sizes is produced that fit together as well as they can.

By contrast, the finely laminated structure of clay shales (*opposite, bottom*) impose no order at all. Readily disintegrating under the action of frost and rain, they invite an entirely random network of criss-crossing break lines.

The aesthetic appeal that links each of these examples, and that of the section of bone below, derives from the attractive combination of geometry and pure chance.

BRANCHA
branching patterns

Branching systems provide an elegant solution where there is a need to access every part of a given area in the most efficient and economical way. That is why these patterns are so widespread in nature and are quite essential to the workings of complex organisms, where the dynamics of efficient distribution (involving energy in one form or another) is at a premium.

The almost universal characteristic of these forms, whether they are conveying water, oxygen, nutrients, or just electrical impulses, is that they consist of a system of finer and finer ramifications, which is another expression of their efficiency.

Branching systems may gather "inwardly" (like the rivers that they so frequently resemble), or they can distribute "outwardly," and they may even support an energy flow in both directions (as is the case in the branching systems of lightning strikes).

Opposite we see a Purkinye cell, a single brain neuron, and below two nautical brancha: Irish moss seaweed and sea fir.

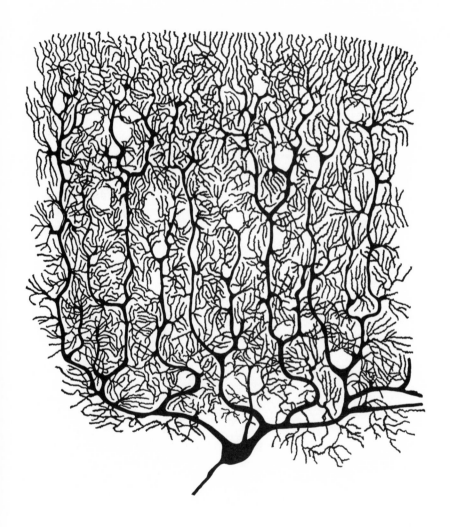

BRECHIA
breaking and separating

This is a class of *li* that is primarily associated with marble and other decorative stones.

The aesthetic appeal of such cut-and-polished panels owes much to their rich coloration, and also to the impression they convey of frozen activity, which seems to present a snapshot of intense formative processes from the remote geological past.

Metamorphic rocks of this kind have indeed acquired their crystalline form under conditions of extreme heat and pressure.

Their characteristic "brechiated" and "veined" forms indicate that an original, static mass has been penetrated and broken apart by a powerful, intrusive flow.

Typically, the brechiated components do not move very far, but remain as a cluster of islands, threaded about by a streaming network of channels—serpentine (*below*) and steatite (*opposite*).

As with most *li,* these formations offer numerable and diverse analogies—not least with the notion of cultural penetration and revivification through the influx of new ideas.

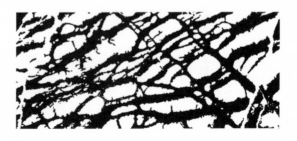

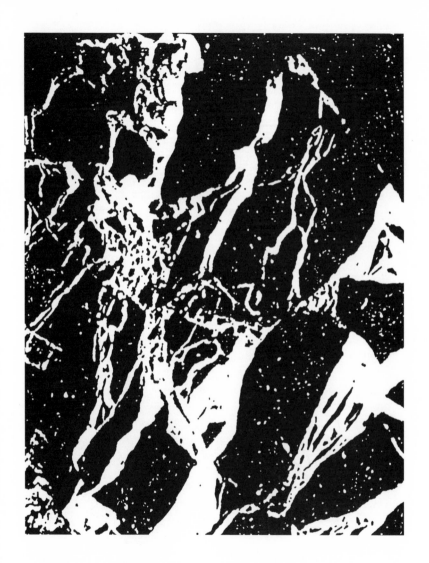

CELLULAR
basic organic arrangements

All organisms are made up of cells; they comprise the basic structural units of all organic matter.

Although they appear in a bewildering variety of forms, it is possible to make general statements on their nature.

By virtue of their highly regulated internal structure, cells are endowed with an almost crystalline orderliness, but they are also possessed by a dynamic sense of purpose; they are functional entities, working to supply the organism with all of its needs.

The dynamic organization within cells is paralleled by their relations with their neighbors, with whom they usually coexist in very close proximity.

In essence, all living things are symbiotic at heart; the properties of complex organisms are an expression of the separate activities of their component cells, and each cell lives in the specific environment created by this association.

The *li* of cell structures, like the vascular cambium (the active layer of tree cells) of *Juglans* and *Robinia* (*opposite and below*), indicate this elegant and sensitive division of space and function.

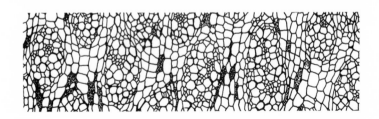

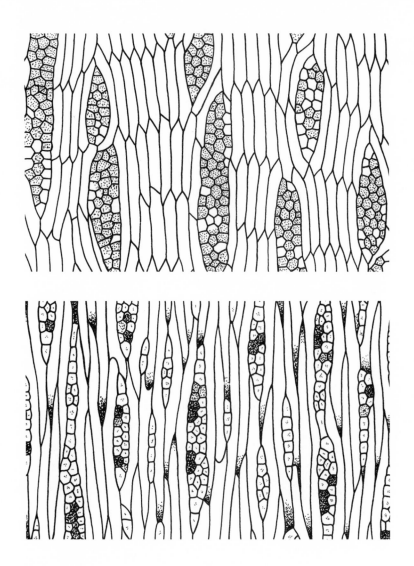

Concentra
propagation around centers

Apart from their rich coloring, what is the real appeal of semiprecious stones such as agate (*opposite, top left and bottom*) and malachite (*opposite, top right*)? To say that they are inherently beautiful simply begs the question: Why *do* we find such formations attractive?

The intricate, concentric configurations found in these and other stones derive from gradual and complex processes of diffusion, water seepage, and crystalline activity.

Much of the color and the tonal variation are the result of impurities in the accumulating layers and of competition between rival centers of crystal growth.

Most important, the details of all this activity (which has occurred in a time and place remote from us) are revealed in sharply defined forms.

The waves and spirals (*below*) resulting from the diffusion process of the famous Belousov-Zhabotinsky chemical reaction, content to merrily oscillate away in their dish without any reference to ourselves, are engaging for much the same reason.

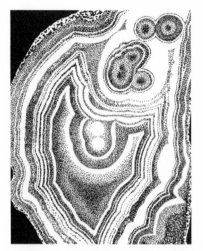

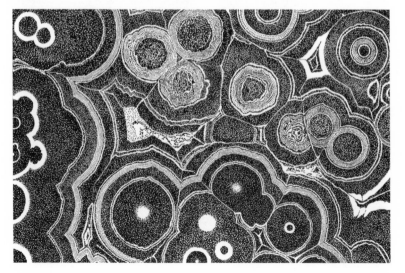

CONTORNARE
physiographical configurations

This category of *li* includes all manner of topographical boundaries and contours. They are obviously among the largest of all formations, but they indicate the general scalelessness of *li* like no other.

Many of the contours (*below*) and textures in the images of landmasses viewed from space seem so reminiscent of those on our own level of existence that it can be difficult to grasp the great differences of scale involved. Some formations can indeed look positively organic, evoking the forms of plants or the markings of animals.

Where there are features with well-delineated boundaries, as in these images of coastlines (*opposite, top*) and snowlines (*opposite, bottom*), the topography is more strongly revealed, clearly exposing the imprint of the active formation processes involved.

This indicates another aspect of *li:* their essentially uncovenanted nature—they are neutral. Like the forces that create landscapes, they are intimately involved in the processes of creation and destruction but are neither innately creative nor destructive in themselves—they just are.

CRACKLE
shrinkage patterns and crazing networks

It surely says a great deal about their respective worldviews that the Chinese have long placed an aesthetic value on the craze patterns on their pottery, whereas in the West they were usually regarded as an unintended flaw, a nuisance.

The hairline cracks in ceramics (*opposite, top*) occur as a result of the shrinkage differential between the glaze and the body of the pottery. They are closely related to the cracks that appear in parched earth (*below*), and to those in dried-out paints and gels (*opposite, bottom*).

All can be seen as force lines, pathways created by the release of a buildup of stress, and it was this aspect that made them so engaging to the Chinese, with their cultural awareness of invisible energies.

With their hierarchies of greater and lesser highways (which correspond to the sequence of their formation) these *li* have many, many correspondences, including the Earth's tectonic system and the street plans of cities.

FILICES
fernlike formations

For those who live in cold northerly latitudes the delicate frosty traceries that occasionally decorate their windowpanes (*see below*) can be one of the more surprising consolations of a hard winter.

It can also be surprising when we first realize that the luxuriant outgrowths occasionally found on rocks, such as those opposite, are not fossils of some ancient plant, indeed, are not of organic origin at all.

The prosaic explanation for these and other similar formations is that they are crystalline; their elongated dendritic crystal forms grow for a distance before forming fresh nuclei and new outgrowths.

There is often an almost magical quality to these figures; they can appear to be positively imbued with life and purposefulness. They exemplify the notion that *li* "derive from a natural and inescapable law of affairs and things that arise directly out of the nature of the universe" (Chhen Shun c. 1200 C.E.).

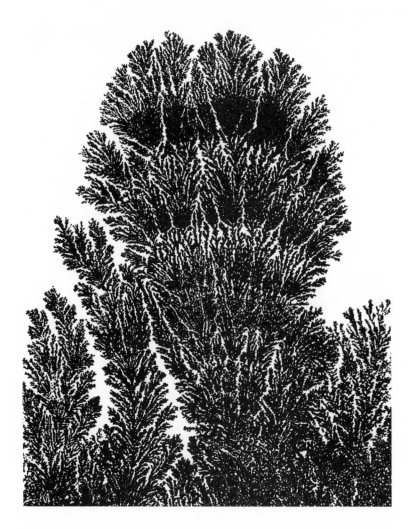

FRACTURE
cracks in elastic materials

Although they bear a superficial similarity, the pattern of cracks appearing in asphalt paving (*see below*) are of a different order from those illustrated on pages 20 and 21.

The difference lies in the nature of the materials involved: asphalt is essentially elastic, pottery glazes are inelastic. Among other characteristics the latter has far more right-angle joints.

The fissures found in many tree barks are some of the more familiar examples of fractures in an elastic medium. These fissures arise as a result of tension on the outer bark of the tree, which is caused by the growth of its inner core; different species have evolved different strategies to resolve this difficulty.

The longitudinal cracks in pines (*opposite, top left*) typically create cells, not unlike those formed in asphalt. By contrast the sweet chestnut (*opposite, top right*) directs the expansive forces into an elegant, gently helical system of furrows.

Each species has its distinctive *li* (or, one might say, each *li* produces its distinctive species). Mature trees are also affected by the piling up of new bark material, which deepens and emphasizes the cracks, as in the oaks (*opposite, bottom*).

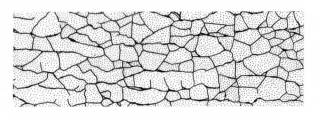

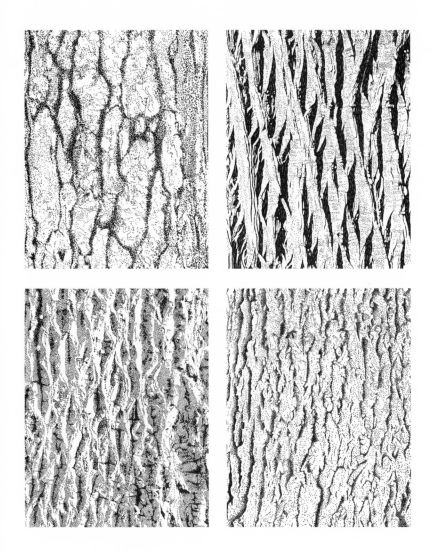

LABYRINTHINE
natural maze patterns

Many civilizations, both ancient and modern, have created mazes and labyrinths. The motives for building them are not always clear, and it seems probable that different cultures had different uses for them. But what general impulse is it that they satisfy? Do they in some way reflect the workings of our mental and psychic processes, or is their complexity simply a measure of the limits of our intellectual capacity?

Whatever the basis of our abiding fascination with these artifacts, it is the case that mind-teasing patterns also appear in nature and that they can arise from a whole variety of different causes.

Interestingly, the common causal factor often appears to be the dislocation of a more complete pattern, as with fragmented liquid convection rolls (*below*) and magnetic maze-domain patterns from a silicon-iron polished crystal (*opposite, top*). At least these are negotiable; succinic acid crystals (*opposite, bottom*) present a disturbed array of bewildering complexity.

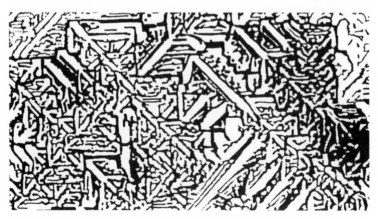

LICHENIFORM
lichen and lichenlike growth

The flat encrustations of lichen, that ancient colonizer of rocks, trees, and old walls, are familiar to all. Its formative habits are appropriately simple for such a humble plant, but the general, cumulative effect of successive layers of growth can be delightfully complex (*below*).

Emanating from centers, receding from these centers, occasionally forming boundaries with its neighbors, and frequently overlapping them, lichen are the beautifiers of bare, inhospitable surfaces.

Some aspects of lichen-growth patterns (the fractalline advance and deference toward adjacent perimeters) are reflected in certain chemical reactions, as in the formation of crusts of gallium oxide (*opposite*).

It is interesting here that a fracturelike network is created.

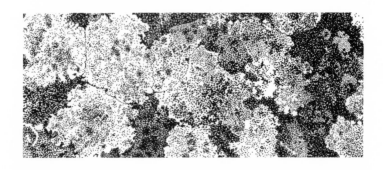

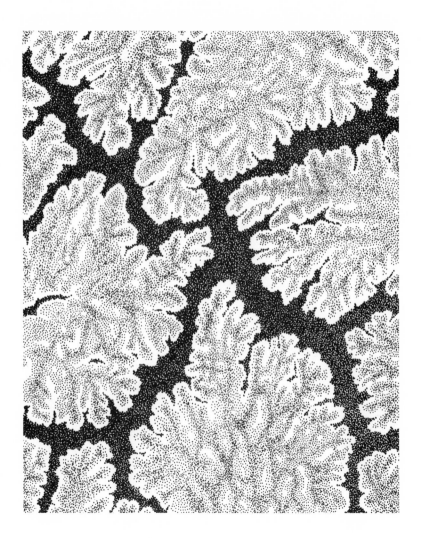

NUBILOUS
cloud and cloudlike formations

Clouds, with their amorphous boundaries and turbulent structures, are justifiably considered to be at the very edge of possessing a definable form.

There is, nevertheless, an internationally agreed system of identification that places clouds into various families, genera, and species (the controversial history of which makes an interesting study in itself). A familiar cloud form is shown below, *altocumulus*, or "mackerel sky."

Nebulous though they may be, clouds clearly show the imprint of the forces working upon them; they express distinctive *li*.

These cloud *li*, whether billowing, laminal, or fractured, naturally have their resonances in other mediums—those opposite are from the vegetable and mineral kingdoms, the xylem of *Salix nigra*, black willow (*top*), folded Lewisian gneiss (*center*), and metal impurities in jaspar (*bottom*).

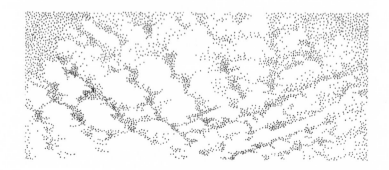

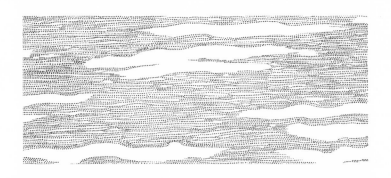

PHYLLOTAXY
dynamic spirality

The leaf-distribution system of plants is determined by a strict mathematical progression (the famous Fibonacci series), but where this arrangement is closely confined, as in the cabbage, the leaves jostle with each other to create a complex, chaotic mass.

A succession of slices through a cabbage (which is essentially a greatly enlarged terminal bud) reveals how the underlying geometrical sequence is disrupted by the differential growth of individual leaves.

This progression, from a relatively orderly to a far more fluid formation, can be seen as a paradigm for the interaction between the two great principles of form and energy—and of the resultant complexity.

That such a humble example can invoke such an elevated analogy at all is, of course, indicative of the underlying reasons for the enduring aesthetic appeal of *li*.

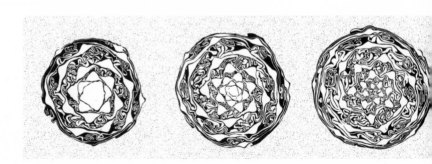

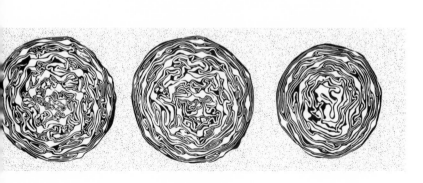

POLYGONAL
geometric quasi regularity

The diagram below is an outline of a section of the basalt columns at The Devil's Postpile, in California.

This amazing formation, like others of its kind, was formed as laval basalt cooled and crystallized, creating a "3-connected" joint network.

In an ideal, perfectly homogenous material this would produce a pattern of regular hexagonal columns. However, the realization of a geometric ideal is rare here, as it is elsewhere in nature.

A similar, but rather less regular, pattern occurs in a film of soap bubbles between glass plates (*opposite, bottom left*), and in arrangements of tissue-forming parenchymatous plant cells (*opposite, bottom right*).

A section of the stalk of a dead nettle (*opposite, top*) reveals a curiously irregular cell structure that appears to vacillate between the various possible systems of orderly close packing.

In fact, the appearance of such disturbances is the distinguishing mark of *li;* perfectly ordered arrays belong to the domain of pure symmetry.

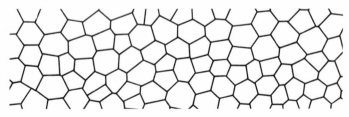

RETIFORM
netlike formations

Latticelike divisions, reticulations, are a feature of many *li*, and there is an obvious correspondence between these examples, which are all drawn from the insect world, and a wide range of other natural forms.

The insect wing below, with its cells arranged in an elegant, dynamic tension, has clear similarities with the soap-film patterns of the previous page—and shares with them the same economic principles of optimal form.

With the katydid (*Paraphidia; opposite, top*) we encounter the fascinating and complicated subject of *li* as camouflage. As is often the case, the aim of this disguise is both specific and general; the concealing pattern resembles that of a decaying leaf, but may also help the creature to blend in with its background in a more general way (*see appendix, pages 52–58*).

The lower image opposite is not, as may be assumed at first glance, the web of a some macramé-minded spider, but part of the wing casing of another variety of grasshopper (*Taeniopoda maxima*).

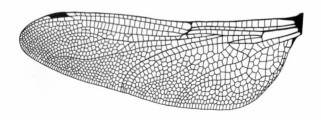

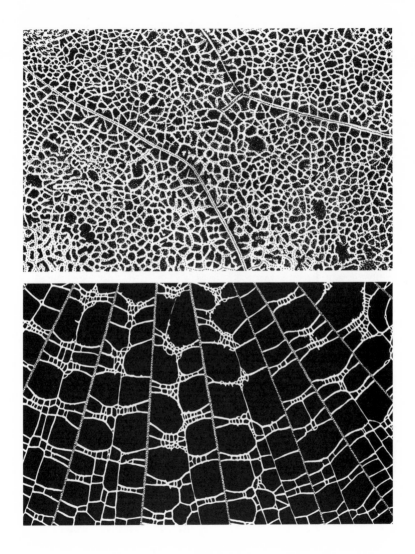

RIVAS
river drainage systems

Rivers and their networks of tributaries present one of the most familiar of all *li* images, forms that have obvious affinities with many other liquid-channeling complexes—particularly those of the essential circulatory systems of animals and plants.

It is clear that these are representations not of mere conduits, but portray the most active part of the Earth's hydrological cycle, and as such are important energy distribution patterns.

With the formation of the river systems there is a certain paradox of priority; rivers shape the landscape, and the landscape contains the river. But, like the chicken and the egg riddle, it is unclear which came first. There is, in fact, a distinct sense of the inevitability of such forms, a view that if carried through credits them with the almost Platonic noumenon of preexistence.

In more conventional terms, drainage patterns are classified into various ideal systems: dendritic, trellis, pinnate, parallel, radial, and rectangular (*opposite, bottom*). The picture opposite, top shows pinnate drainage patterns in Sumatra. Below we see a complex anastomosing channel system in Bangladesh.

RIPPLES AND DUNES
wind and current patterns in sand

Sand is itself inert, but wherever it is found it bears the imprint of the forces that have been working upon it.

These forms, including the engaging variety of ripples on the shore and the endless, impressive dunes of the desert, appear to have an existence of their own, governed by laws of their own.

Although the material of which they are composed is being constantly reworked, these sand *li* themselves are relatively constant.

Whether they are aeolian (wind-formed) or, in the case of shore patterns, created by the compound effects of currents, wind, and tide, their character is strongly influenced by particular local conditions.

As a result they exist in infinite variety and perfectly demonstrate the principle that simple causes can produce irreducibly complex effects.

The abiding characteristic of these formations, common to many *li*, is the sense of an overall order, showing clear repetitive features, but lacking a strict periodicity.

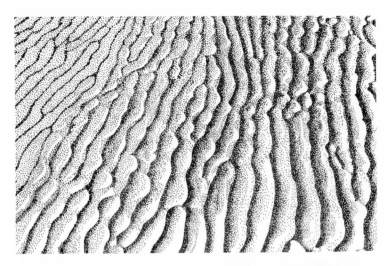

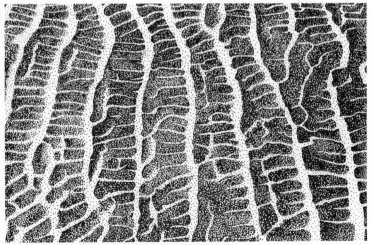

41

TRIGONS
triangular formations

Triangles are rare in nature, but the surfaces of diamonds, when microscopically examined, reveal a system of contours in which regular triangles are the most prominent motif (*opposite*).

Diamonds, of course, are the very symbol of crystalline perfection, and crystals in general are by far the most symmetrical objects in nature, with millions upon millions of identical atoms held in fixed positions within a predetermined lattice structure.

It is the case, however, that even the most flawless of these arrays is teeming with minute defects, arising from the dislocation of atoms—and these are the ultimate source of diamonds' triangular formations.

In this aspect of nature, as in every other, energetic forces engage with formative principles to create a distinctive *li*.

The marvelous cone shells of tropical seas, like the *Conus marmoreus* and *Conus textile* (*below*), are another rich source of triangular *li*.

VARIEGATUS
spots, speckles, and scrawls

One of the most universal and familiar formative processes is the tendency to form clusters of all kinds.

In the inorganic world this is typified by the sort of variegated, clumping effects that we observe in magmatic rocks (*below, left*), and the fractalline tremas of surfactant films (*below, right*).

In busy natural environments, variegated forms of one kind or another provide much of the background, so it is hardly surprising that the creatures that inhabit them adopt similar configurations as cryptic markings.

The skin patterns of various frogs and toads (*opposite, top row*) provide a particularly rich field of these sort of multiform *li*. Equally intriguing formations can be found in subjects as diverse as the wing cases of insects (Goliath beetle, *opposite, center left*) and seashells (Tiger cowrie, *opposite, center right*).

The eggs of most ground-nesting birds also need to blend with their surroundings, so these provide another, extensive range of mottled *li*. The bottom row opposite shows Oystercatcher, Common Murre, Razorbill, and Black Guillemot eggs.

 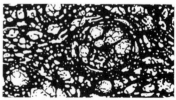

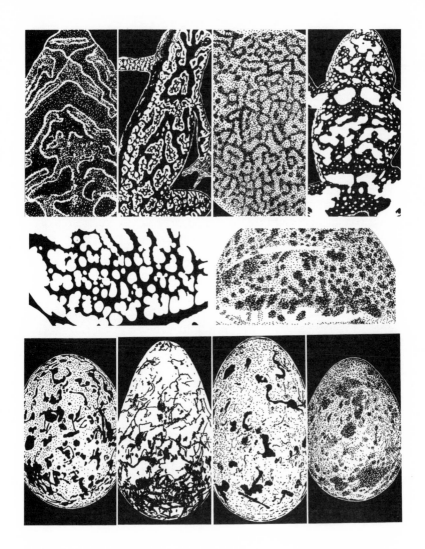

VASCULUM
leaf vascular patterns

Leaves of all kinds are involved in the world's single most important chemical transaction: converting the Sun's energy into food. Their veining systems play a critical role in this process, supplying water and mineral salts to every part of the leaf and efficiently removing the elaborate food compounds created by photosynthesis. In addition, vascular systems provide a supporting skeleton for the leaf, so they are a proficient plumbing system and superstructure combined. These are very attractive forms, in all their great variety and subtlety.

There are obvious similarities between these plant circulatory systems and the venous and nervous systems of animals—but they are also strangely evocative of river drainage complexes. The common factor in all of these processes is, as ever, the transfer of energy; these delicate forms should also be seen as energy pathways.

The veins shown here are from *Ficus carica* (*below*), *Liriodendron tulipifera* (*opposite, top*) and *Quiina acutangula* (*opposite, bottom*).

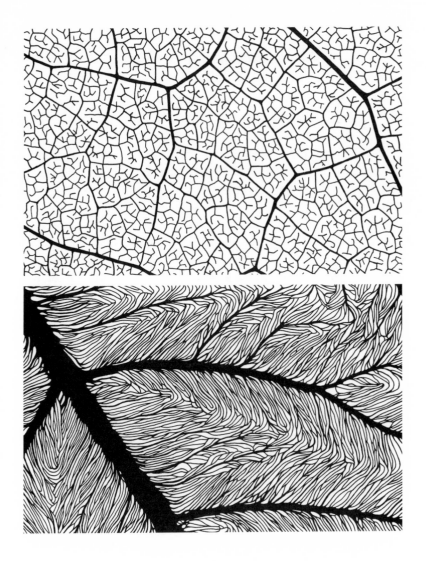

VERMICULATE
wormlike arrangements

These intriguing figures, found among a range of quite different processes, arise essentially from the interplay of mutually repellant domains.

These forms are of great interest to physicists because of their relation to turbulence, a common condition the inner nature of which, however, is one of the great remaining mysteries of physical science. They are some of the most dynamic of all *li*, often appearing fleetingly and undergoing constant and rapid change.

When fluid is vigorously heated, for instance, many small spirals appear that mingle and react with each other to form a spiral-defect chaos (*opposite, top left*). A similar effect occurs as a result of the interaction of magnetocrystalline energies on a garnet film, so-called magnetic domain patterns (*below,* and *opposite, top right*). In an entirely different setting, a whole range of serpentine blobs known as Langmuir films can appear when surfactant films are compressed on a water surface (*opposite, bottom left*).

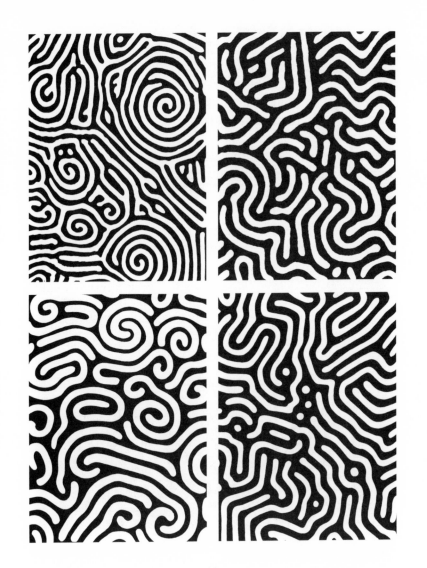

Viscous Maculae
release patterns

It seems appropriate to end this survey of *li* with a category that can easily be investigated by the readers themselves, with a minimum of equipment.

The designs opposite and below were formed simply by prying apart two small sheets of glass with ink spread between them.

The resultant configurations, essentially extended blots, are often surprisingly intricate. Evocative of forms as diverse as marine vegetation or flames, they are a perfect demonstration of the sort of complexity that can arise from the most basic of initial conditions.

That complexity can arise almost of itself, and that there is a connectedness between every part of the self-created cosmos is, of course, a central tenet of the Chinese philosophy of Taoism. The *Tao te Ching*, its principle text, affirms this:

> *"The Way (Tao) is a thing incommensurable, impalpable,*
> *Yet latent in it are forms (Li)."*

Appendix
LI IN ANIMAL MARKINGS

One of the more intriguing aspects of *li* lies in the unlikely similitudes that occur as a result of entirely unrelated formative processes—say, between the bands created by sand dunes and those on a Zebra's skin, or the network of cracks in dried mud and the markings of a giraffe.

Animal markings are a rich source of *li*, and there are many such comparisons. Of course, an animal's visual aspect is intrinsic to a whole range of functions: it can camouflage for defense or attack; it can warn off and confuse; and it is usually involved in the business of sexual attraction. Frequently these different functions are combined.

Quite often, though, an animal's appearance cannot be entirely explained in these purely functional terms; the way a creature looks, and the particular *li* that it adopts, is an essential part of its being.

In the final analysis *li* are part of the ultimate mystery of creation.

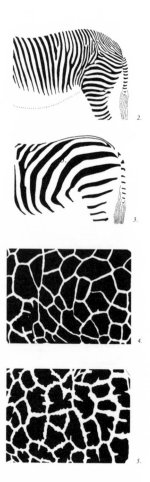

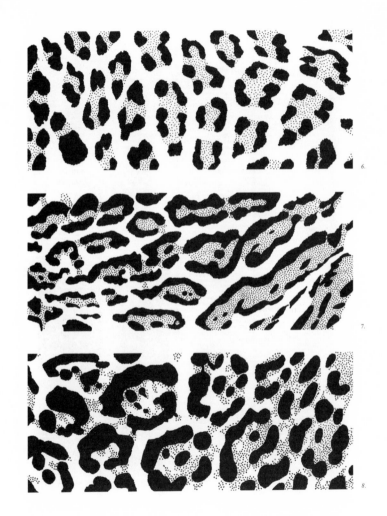

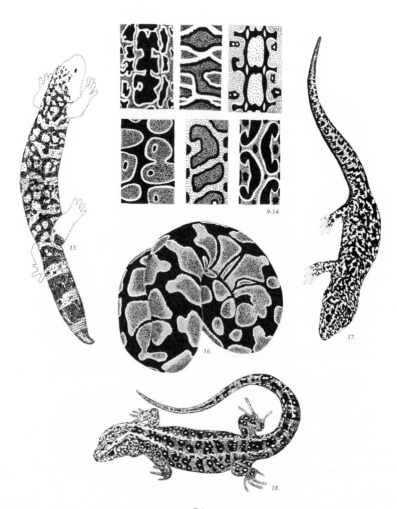

PISCES

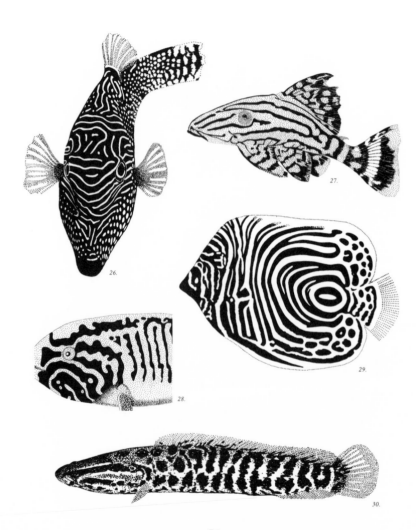

CONCHA

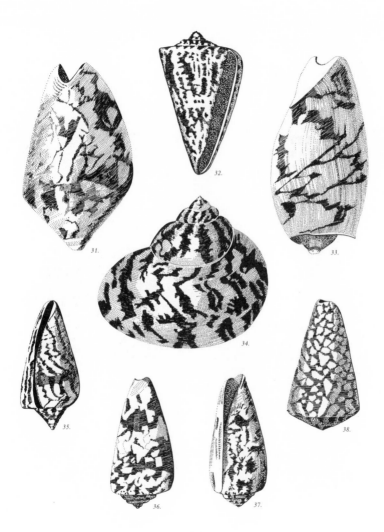

MOLLUSCA

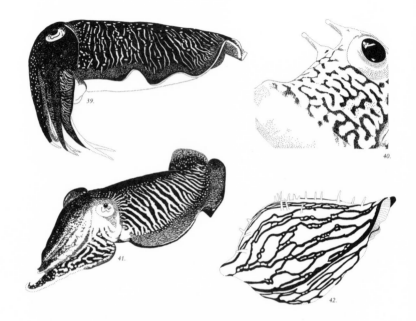

LIST OF APPENDIX ILLUSTRATIONS

1. Baby tapir 2. Common zebra 3. Grévy's zebra 4. Giraffe, Camelopardalis reticulata
5. Giraffe, camelopardalis tippelskirchi 6. Leopard 7. Ocelot 8. Jaguar
9. Python molurus 10. Natrix fasciata 11. Constrictor constrictor 12. Python regius
13. Ophibolus doliatus triangulus 14. Bothrops alternatus 15. Gila monster
16. Royal python, curled 17. Night lizard 18. Sand lizard 19. Poison-arrow frog
20. Patterned tree frog 21. Yellow-banded poison-arrow frog 22. Fire salamander
23. Dendrobatus tinctorius 24. African tree frog 25. Hyperolius marmoratus
26. Sharp-nosed puffer 27. Panaque nigrolineatus 28. Cuckoo wrasse
29. Emperor butterfly fish 30. Snakehead 31. Noble volute 32. Lettered cone
33. Elephant-snout volute 34. Magpie shell, Cittarium pica 35. Spectrum cone
36., 37. Striated cone 38. Marbled cone 39. Giant cuttlefish 40. Sea slug
41. Cuttlefish 42. Mollusc "cloak"